The Birth and Death
of the Cosmos

The Birth and Death of the Cosmos

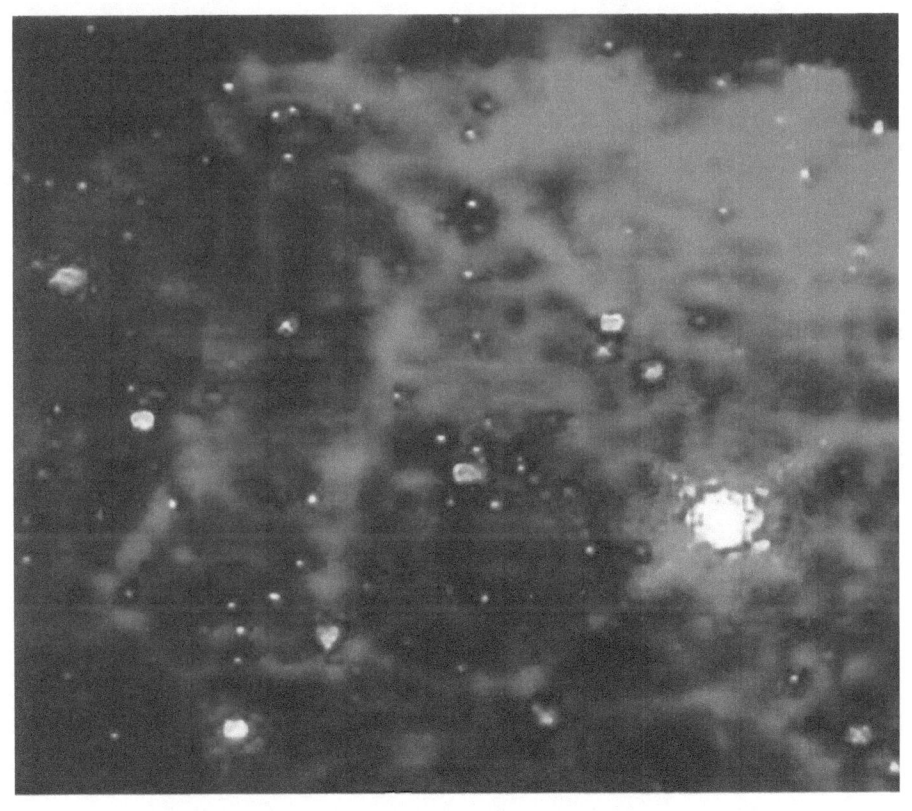

Dr. Terrance A. Austin

To order additional copies of this book, contact:
Xlibris Corporation
1-888-795-4274
www.Xlibris.com
Orders@Xlibris.com
60584

Contents

Preface

OFTEN PEOPLE THROUGHOUT THE WORLD, HAVE WONDERED WHERE THEY CAME FROM AND WHERE THEY WERE GOING. AS WE BECAME MORE EDUCATED THROUGH TIME, OUR BELIEFS RAN THE GAUNTLET FROM THE IDEA THAT WE WERE ALL ALONE IN THE UNIVERSE, RANGING FROM A BEGINNING OR HELD THAT THE EARTH WAS AT THE CENTER OF THE UNIVERSE AND OUR GALAXY WAS THE ONLY GALAXY. WE PROGRESSED TO THE POINT THAT ALL THINGS DID NOT REVOLVE AROUND THE EARTH BECAUSE THEY REVOLVED AROUND THE SUN. WE THEN DISCOVERED THAT THE ORBITS OF THE PLANETS WERE ELIPTICAL, NOT CIRCULAR. WE FINALLY DEVELOPED SCIENTIFICALLY SO THAT WE COULD SEE THAT THERE WERE MANY GALAXIES IN THE UNIVERSE, HENCE WE WERE NOT ALONE AFTER ALL, BUT MAY HAVE NEIGHBORS, EVEN THOUGH THEY WERE FAR AWAY. IT WAS THAT REALIZATION THAT BROUGHT PEOPLE TO THINK THAT PERHAPS WE WERE NOT ALONE IN THE UNIVERSE. THIS DOCUMENT DELVES INTO THE BIRTH OF THE UNIVERSE, NOT THE EVOLUTION OF MANKIND UPON THE EARTH, BUT THE UNIVERSE'S BEGINNING, GROWTH AND ENDING WITH A SIDELINE AS TO HOW WELL WE HUMANS WILL FARE IN THE PROCESS. YOU MAY BE SURPRISED, ALTHOUGH SUPRISE IN NOT THE PURPOSE OF THIS DOCUMENT, TRUTH VIA SCIENTIFIC FACT IS. THE AUTHOR HAS INTENTIONALLY OMITTED MANY MATH FACTS THAT WOULD FURTHER PROVE HIS POINT BUT HAS CONTAINED ENOUGH MATH TO PLACE REALITY INTO THE DOCUMENT.

Chapter 1

Introduction

There came a time in the age of man when he began to think in the abstract. That time has not been determined but let's simply draw intersecting lines from the Renaissance, the birth of Christianity and the discovery of Einstein's Relativity and use that intersection as a starting point. During this intense time of thought, mankind seemed to be divided into two camps pertinent to the creation of the universe. One faction said that the universe has always been and will forever be. This group meant that the universe was never created at all but a large mass of infinitely dense matter packed together that went bang, expanded till the pull of gravity stopped the expansion then gravity pulled it all together again where it went bang again. This constant exploding, has of course, lasted forever. The opposing viewpoint said that God created the matter for the universe and that God caused it to go bang for the Universe in total at his will. They concluded that God created the

matter. Neither side would give on their viewpoint and, hence, The argument

began and is still going on. We are going to delve into this issue via several

sidetracks wherein we will look at the logic of these sideline events. We will

look into the absolute distances involved in this arguments as well as the

logic or lack of it pertinent to these issues and arguments.

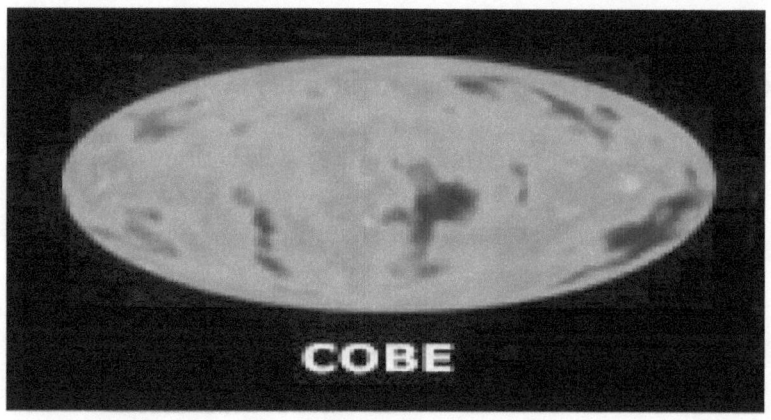

Universe Fig 1

The Wilkinson Microwave Anisotropy Probe (WMAP) team has made

the first detailed full-sky map of the oldest light in the universe. It is

a "baby picture" of the universe. Colors indicate "warmer" (red) and

"cooler"(blue) spots. The oval shape is a projection to display the whole

sky; similar to the way the globe of the earth can be projected as an oval.

The microwave light captured in this picture is from 379,000 years after the Big Bang, over 13 billion years ago: The image is portrayed as an oval, the same as one can picture Earth on maps in order to avoid distortion dirbe123_2 p6dec

Fig 2
The Milky Way (side view)

First let's look at the distances involved in the whole universe so we can get a feel for these astronomical numbers. Light travels at 186,000 miles per second. In one minute light travels 11,116,000 miles. In one hour light travels 669,600,000 miles, during one day light goes 16,070,400,000 miles in a day and 5,865,696,000,000 miles in a year. That's right the figure reads near 6 trillion miles a year. Scientists say that the big bang occurred about 15 billion years ago which is now gives us a radius of 90 trillion-billion miles for the whole universe if the universe runs at the speed of light. Now in order to more clearly envision these distances we must embark into where we live now, where we think we must go and why, how will we get there and how will we propel our spacecraft? Along the way we are going to encounter some laws and what we now consider scientific facts that we will put into dispute and explain why they are being disputed. To do this we will look briefly at Einstein's relativity, we must also look at our universe as we now see it. Our universe as it might have to be, plus a few side visits to some facts of science that don't seem to hold up to common sense. Even though we will take these side trips we will always return to our original destination of the origins of the universe.

Chapter 2

Distances

The distances previously mentioned are almost too large to fathom. Scientists use scientific notation to address these very large numbers; hence a billion becomes 1 times 10 to the ninth power. This may be easier for scientists in grasp but in reality it is nothing but a very large number. For example the universe which is currently thought to be 15 billion light years old spans a radius of 90 trillion, billion miles. This figure comes from what science now believes is the time of the "Big Bang" which we are fairly sure occurred 15 billion years ago times the distance that light travels in a year. The problem lies in the fact that neither side has any real facts to back up their case. There is only almost blind faith on the side of gravity controlling the infinite series of big bangs. Science on the other hand has tried many theories to fill in the gap between the two thoughts. You see the problem lies in the fact that when we measure the amount of matter in the universe we can only account for about 10% of the total matter necessary to make a big bang. For years science said that the missing matter consisted of dark matter scattered among the universe but measurement have never come close to discovering enough dark matter (matter that doesn't reflect nor emit light but instead alters the relationship between gravity and the force expanding our universe by helping the original force of the Big Bang by repelling gravitational forces). The interesting point that scientists make is that the matter consisting of the Big Bang was at an infinitely high temp and they were infinitely dense. This is the premise one must take on faith.

The good news from NASA's Hubble Space Telescope is that Einstein was right—maybe. A strange form of energy called "dark energy" is looking a little more like the repulsive force that Einstein theorized in an attempt to balance the universe against its own gravity. Even if Einstein turns out to be wrong, the universe's dark energy probably won't destroy the universe any sooner than about 30 billion years from now, say Hubble researchers. "Right now we're about twice as confident than before that Einstein's cosmological constant is real, or at least dark energy does not appear to be changing fast enough (if at all) to cause an end to the universe anytime soon," says Adam Riess of the Space Telescope Science Institute, Baltimore. Riess used Hubble to find nature's own "weapons of mass destruction"—very distant supernovae that exploded when the universe was less than half its current age. The apparent brightness of a certain type of supernova gives cosmologists a way to measure the expansion rate of the universe at different times in the past. Riess and his team joined efforts with the Great Observatories Origins Deep Survey (GOODS) program, the largest deep galaxy survey attempted by Hubble to date, to turn the Space Telescope into a supernova search engine on an unprecedented scale. In the process, they discovered 42 new supernovae in the GOODS area, including 6 of the 7 most distant known. Cosmologists understand almost nothing about dark energy even though it appears to comprise about 70 percent of the universe. They are desperately seeking to uncover its two most fundamental properties: its strength and its permanence. In a paper to be published in the Astrophysical Journal, Riess and his collaborators have made the first meaningful measurement of the second property, its permanence.

Currently, there are two leading interpretations for the dark energy as well as many more exotic possibilities. It could be an energy percolating from empty space as Einstein's theorized "cosmological constant," an interpretation which predicts that dark energy is unchanging and of a prescribed strength. An alternative possibility is that dark energy is associated with a changing energy field dubbed "quintessence." This field would be causing the current acceleration—a milder version of the inflationary episode from which the early universe emerged. When astronomers first realized the universe was accelerating, the conventional wisdom was that it would expand forever. However, until we better understand the nature of dark energy—its properties—other scenarios for the fate of the universe are possible. If the repulsion from dark energy is or becomes stronger than Einstein's prediction, the universe may be torn apart by a future "Big Rip," during which the universe expands so violently that first the galaxies, then the stars, then planets, and finally atoms come unglued in a catastrophic end of time. Currently this idea is very speculative, but being pursued by theorists. At the other extreme, a variable dark energy might fade away and then flip in force such that it pulls the universe together rather then pushing it apart. This would lead to a "big crunch" where the universe ultimately implodes. "This looks like the least likely scenario at present," says Riess. Understanding dark

energy and determining the universe's ultimate fate will require further

observations. Hubble and future space telescopes capable of looking more

than halfway across the universe will be needed to achieve the necessary

precision. The determination of the properties of dark energy has become

the key goal of astronomy and physics today.

Release Date: 12:00PM (EST) February 20, 2004

Release Number: STScI-2004-12

Contact:

Don Savage

NASA Headquarters, Washington, DC

To fill in the missing 90% of the matter necessary to make up the big bang,

hence we gave up on dark matter and pushed on to other possibilities. One

theory said that the universe was filled with sub atomic particles to the degree

that millions or billions of these particles passed through our bodies every

second hence going through the Earth as well. We built a large underground

pool and loaded it with sensors to detect the passing of these particles and

now, after years of watching we have not detected a single particle. This

drops sub atomic particles. Now we must get back to distances and the reasons for comprehending these distances, but as a sideline let me point out that there was an article that stated that the diameter of the matter that constituted the big bang was only 0.080 inches in diameter down to the size of an atom. Right, 80 thousandths of an inch in diameter. The only answer is to accept that the matter within the big bang was infinitely dense and also infinitely hot.

Chapter 3

Galactic Size

Let's take a look at our Galaxy as seen from outside our galaxy (a neat trick as we shall see). * The Milky Way is a gravitationally bound collection of at least 100 billion stars. Our sun is one of those stars and is located about 25,000 light years, (8000 parsecs) from the center of the Milky Way. The Galaxy has three major components: A thin disk consisting of young and intermediate age stars—this disk also contains gas and is actively forming new stars. Dust absorbs blue light more than red light and thus makes the stars look reddish. Our Galaxy has spiral arms in it's disk-these spiral arms are regions of active star formation.

- *A bar of older stars (white in the COBE picture).*

- *An extended dark halo whose "http://map.gsfc.nasa.gov/m_uni/ uni_101matter.html" is unknown. Since the matter in the halo does not consist of luminous stars, it does not show up in the COBE image. The existence of the dark halo is inferred from its gravitational pull on the visible matter. "http://lambda.gsfc.nasa. gov/product/cobe/dirbe_image.cfm."*

Chapter 4

Space Travel

Funny, science fiction has placed the thought in our minds that space travel is not only simple but also inevitable. Thus far in our brief history (I use the word brief because we people as a race has only been around a few million years as opposed to the age of the universe at 15 billion years). We have traveled far in our history already going to the moon and Mars. The Milky Way is a spiral Galaxy and like most galaxies has a giant black hole at its center which keeps all stars in the galaxy revolving around it. If one were to picture this black hole one would see a great deal of star matter being pulled into the black hole. The black hole is one of the primary sources of gravity holding the galaxy together but apparently not a strong enough pull to effect most of the other galaxies, many of which are moving away from us and the center of the universe. To view our galaxy from another point of view the following picture presents a spiral galaxy very similar in nature to our milky way.

Fig 3   Galaxy M51

Another view of similar galaxy is shown in the following picture. The Milky Way is a spiral galaxy with a black hole at its center as most galaxies do. This black hole has enough gravitational pull to hold the galaxy in place despite the galaxies vast size. The vast gravitational pull, however, seems to have little effect on the other galaxies as they are pulling away from us and the center of the universe, More will be said about this later.

Fig 4   A spiral Galaxy

It must be pointed out that our sun exists toward the end of one of the spirals at about 7:00 O'clock in the picture, Hence we are not very far from the edge of the galaxy. This does not infer that intergalactic travel might be at hand. In fact, now that we have pointed out the vastness of the universe, we should talk a bit about how we are going to move around in this universe. (Robotically speaking) as well as having sent spacecraft out of our solar system heading for other galaxies. These are very fine accomplishments but speak nothing about how or why we must leave the Earth. We have considered a few different methods of powering space travel those being as follows: Atomically—we could build a spaceship with baffles on its rear end to shield us from radiation, then start firing atomic bombs behind the baffle and in this manner we could theoretically attain speeds of about 100,000 miles per hour. This is really moving by today's standards since our current rockets reach about 25,000 mph to escape the Earths gravity. Even though 100.000 mph is fast it would take about 143 years at that speed to reach the nearest star. Alpha Centaury which is really a group of three stars spaced such that they appear as one star from the Earth. Proxima is the nearest of the three stars and it is 4.2 light years away from Earth. One of the problems with atomic bomb powered spaceships is that in order to slow down, you must have a baffle in the front of the spaceship so you can explode atomic bombs off in the front of the ship so you can slow down and stop when you get there. talk about radiation, Who is going to volunteer to go place & set off the atom bombs in both the front and the back of the ship.

Ion flow Propulsion

We have conceived of another form of propulsion and that is ion propulsion. You place one big parachute in front if the spaceship to catch mass the free ions emitted by all the stars and you can reach theoretical speeds of 300,000 mph. You must also have a parachute in the back so you can slow down when you get there but the real problem is that it will; still take you 40+ years to get to our nearest star. These two speeds are a far sight from what scientific theory tells you about warp speed wherein one can exceed the speed of light but science fiction tells not of any downside for warp speeds.

Matter-antimatter drive.

We know that when we combine a particle of matter with a particle of anti-matter that an explosion uses up all of the matter and anti-matter. The explosion following the mixture of the two particles is the most efficient explosion, orders of magnitude above the power of the atomic bombs man has made. Enough power from a mixture of matter and antimatter could provide us with the power to travel into space. Look out for the first times man tries to combine these particles lest we lose the whole planet from the result.

Warp speed is a binary progression of speed meaning

warp 1 ----------- speed of light

warp 2 ----------- 2x speed of light

warp 3 ----------- 4x the SOL

warp 4 ----------- 8x SOL

warp 5 ----------- 16 x SOL

warp 6 ----------- 32 x SOL

warp 7 ----------- 64 x SOL

warp 8 ----------- 128 x SOL

warp 9 ----------- 256 x SOL

warp 10---------- 512 x SOL

There is one hic-cup in the theory, however, in that Einstein's Relativity tells you that as you approach the speed of light, mass goes to infinity, with mass being the resistance to further acceleration. In short, nothing in this universe can exceed the speed of light and as far as we know nothing does! To see further into this matter we will quote the following:

Warp Theory

> *In the previous section "wdtheory.html", the theoretical science of `warp drive' was briefly discussed. The methods imposed showed the most attractive issues of the theory, however warp drive has one major drawback. The warp drive, does not agree with known conservation laws and hence defies the laws of physics as understood in the late 20th Century. Here I will present these drawbacks in order to present an equal ended forum on the warp drive.*
>
> *Fact: Warp Drive is only at best a theoretical science.*

The Weak Energy Condition

The initial problem of the warp drive was recognized by Alcubierre in his breakthrough paper [\l "Alb"]. The problem being a weak conservation principle in General Relativity (GR) is broken, known as the Weak Energy Condition (WEC). The formula for a non-trivial space is given by TabVaVb=>0. The problem which arises is that it causes the flat spatial metric within the zero Christoffel space to obtain a positive curvature. Thus violating relativity by a local expansion of spacetime, such an effect was predicted much earlier by Einstein. This effect however, was applied to explain an expanding universe, thus when applied locally contradicts Einstein's invention, which is known as the Cosmological Constant.

Expanding and Contracting Spacetime

The idea of spacetime expanding and contracting is not at all new; it should be noted that once again Einstein predicted such an effect. From the local metric gik acting on a Riemannian surface geometry, the curvature connection of spacetime is derived directly from a metric with a signature of rank 2. Thus the curvature connection would be given in a linear form do to the presence of the metric, hence space would be required to expand and contract from the metrics signature. This expansion and contraction of spacetime is called gravitational radiation, which is assumed to propagate at the speed of light c. The key between this idea and Miguel's proposal is that the Alcubierre paper requires two separate sets of gravitational waves. However, do to the spatial metric proposed by Alcubierre one set of the gravitational waves would be required to defy the WEC.

The Energy Conditions

Violating certain conservation laws is to be expected in GR however; this is a consequence of the geodesic nature of spacetime. No path within such a space can be considered linear in its own accord, and hence at some arbitrary scale is expected to violate certain conservation principles. Such a defiance is commonly associated with another Energy Condition in GR, known as the Strong Energy Condition (SEC). A paper by Van

Den Broeck [\l "br1"], takes a similar notion in principle, since it does not completely remove the "negative energy" condition required by Alcubierre. By shrinking the metric by an amount determined by B2(rs), the violation of the WEC is lowered by some 28 orders of magnitude. However with an energy requirement ranking over a hundred trillion trillion fold, this can hardly be considered a minute and expected perturbation induced by GR. This is a real problem, a problem grave enough to allow many physicists to believe such a problem will never be overcome, and that the warp drive should only exist within the realm of science fiction

Quantum Mechanics

While such a "negative" energy scares most with a background in GR, it does not send chills up the spines of Quantum physicists. Negative energy is required by the laws of probability, because of Hesienberg's Uncertainty principle, where one is expected to find negative energy of some form some where.

"http://www.physics.uoguelph.ca/poisson/mitch/warpdrive. html""http://www.physics.uoguelph

.ca/poisson/mitch/warpdrive.html"

Under certain conditions one may even expect a peak in the negative energy far above the norm, while this is expectable in quantum theory, it plays a much to little role in GR. Currently, there would be no way to obtain the conditions necessary for the minimum negative energy which is required by the model proposed by Ford and Pfenning [\l "fp"] (except for some momentary fluke, not likely to be experienced in GR). These problems are no laughing matter; in fact a second paper by Van Den Broeck [\l "br2"], considers the implications of such conditions. Where I now quote:

"... it is clear that warp bubbles present, enormous practical difficulties, which may never be overcome."—Broeck.

More Practical concerns

While many difficult issues have been addressed, such as the negative energy requirement, some rather simplistic problems have been ignored (or rather left to be assumed by the reader). Such as obtaining the mass necessary to curve spacetime in order to induce the warp drive. From the stand point of GR, this is not an initial concern, however it is for body which obeys conservation laws. An object of several tons can not generate a mass several times its own magnitude through any known scientific process. And if it were possible, one runs into the very severe risk of generating a singularity in spacetime, i.e. a black hole. Again, even under the best of circumstances, it appears that negative energy densities are required for the warp drive Thereby from a physical point of view this makes the warp drive look even more unrealistic. Furthermore, a paper by Olum [\l "Olm"], considers a spatial metric of form:

$$ds2=(1-4t2x2)dt2-4tx(1-t2)dxdt+(1-t2)dx2.$$

From the above equation, one may have a space which appears to yield superluminal travel. However, the apparent faster than light motion is just an allusion which appears upon an observers orientation to the metric. The velocity of an object within this metric thus never travels faster than light, at first this may appear to validate the premise of the warp drive. However, it is shown that once again, for apparent FTL motion to occur negative energy densities are required.

References

[1] Alcubierre M. *The Warp Drive: Hyper-Fast Travel Within General Relativity.*

Class.Quant.Grav. 11 (1994), L73-77. available: "http:// xxx.lanl.gov/abs/gr-qc/0009013"

[2] Broeck C. *A `warp drive' with more reasonable total energy requirements.*

Class.Quant.Grav. 16 (1999) 3973-79 available: "http:// xxx.lanl.gov/abs/gr-qc/9905084"

[3] Pfenning M. and Ford L. *The unphysical nature of "Warp Drive." Class.Quant.Grav. 14*

(1997) 1743-51 available: "http://xxx.lanl.gov/abs/ gr-qc/9702026"

[4] Broeck C. *On the (im)possiblity of warp bubbles.* "http:// xxx.lanl.gov/abs/gr-qc/9906050"

[5] Olum K. *Superluminal travel requires negative energies.*

Phys.Rev.Lett. 81 (1998) 3567-70

"http://xxx.lanl.gov/abs/gr-qc/9805003"

As is the case with so many scientific papers, there is much to be said for what is said between the lines. This is a good case of reading between the lines. Unstated is the proposal that conventional fuels do not provide enough power to allow us to reach FTL (Faster than light) speeds. It appears that we must use a matter-anti matter collision to get enough power to reach our goal. Such a collision converts all that is matter into energy, and is orders of magnitude more powerful than any atomic device that currently exists. We don't have a matter-antimatter combining device, and Lord help the world when someone first tries to mix the two for the energy created could cost the world its life. Let me digress for a moment. Our sun is 5 billion years old now and will burn for another 5 billion years. When the time comes that our sun starts running out of hydrogen a few things will happen, The sun will not be able to maintain its current shape and as the hydrogen depletes, the sun will start spread out, then collapse to a small portion of its current size. After it collapses the remaining hydrogen will re-ignite with a fury and the corona will expand outward and cover the space now occupied by our first four planets. This coverage included the Earth, which is in the third orbital position. The corona of the newly formed sun will be over 1 million degrees and all life on the Earth will cease to exist, all water will be boiled off the Earth's surface and the steam will be cast into space. The Earth will be simply a moving rock. It could be worse, for if the corona heats up the Earths core; the expanding gasses from the core could explode this planet into a (technical word) kajillion pieces. In short, this is the definition of Armageddon for it spells the end of the Earth, not just another disaster such as a big meteor striking the Earth. What has gone practically unsaid in this about warp speeds is the impression left by science fiction that at warp speeds we can tour the galaxy or the universe at speeds wherein the stars are whizzing by our front window in a blur. This always appears to be happening at warp 2-4. In reality, at warp 4, it would take one year to reach our nearest star, That is hardly having stars whiz by us. We would creep up on them as though in slow motion even though we were traveling at warp speeds. Let's face facts. If our Galaxy is 200,000 light years in diameter, It would take us 50,000 years to go across the galaxy at warp 4

Even at warp 10 it would take 40,000 years to cross the galaxy. This is hardly a practical method for trans galaxy travel. Another problem that rears up pertinent to warp theory is the fact that your ship must be wrapped in a warp bubble in order to confuse space into thinking you are not really there so you can exceed the speed of light. The amount of energy needed to envelope a ship in a warp bubble is so vast that all the energy in the universe (including all of the stars energy) will not be enough to make one warp bubble big enough to envelope one proton from one atom, much less a whole atom or a spaceship. We must come to the conclusion that warp speeds are beyond us at this time and hope that a similar answer will be found by our progeny. But make no mistake, mankind will have to find a way to leave the Earth and find a new home simply because the Earth will no longer support life if it still exist. Further we must probably travel to a far away star system containing an M class planet (like the Earth so it can sustain life). Solving the problem of how we get there is beyond us now but I have no doubt that we will find a solution. We must get this back to what we aimed at in the beginning.

Chapter 5
FTL

Faster than light we are going to discuss things in our universe that do go faster than light. We will do so to set up the premise that FTL things do exist. We are also going to enter a discussion of what things are within the universe and things outside the universe. All the math that we attribute to our universe have been shown and subsequently proven by Einstein & others who have used his work to complete their own the writer included. Black holes regularly emit particles from within, however, due to the extremely strong gravitational pull from the black hole, the particles must travel at greater than light speeds in order to escape the event horizon of the black hole.

Author: janette l gubala

What is beyond space?

Author: asmith

Nothing! Either space goes on forever (is infinite) or it comes back around in some kind of closed loop, but the way we understand space right now, it is impossible for it to have any edges, and so there is no direction you could point and say "50 yards in that direction space ends". Since there are not any ends, there is not really any way to understand what "beyond" means. But there could be other things that "exist" that are somehow outside our own universe—parallel universes!

"http://www.newton.dep.anl.gov/ askasci/astron98.htm"	"http://www.newton.dep.anl.gov/ archive.htm"
"http://www.newton.dep.anl.gov"	"http://www.newton.dep.anl.gov/ aasquest.htm"

"telnet://newton.dep.anl.gov" is an electronic community for Science, Math, and Computer Science K-12 Educators.

Argonne National Laboratory, Division of Educational Programs, Harold Myron, Ph.D.,

Division Director

A speck of debris is all it takes to knock out spacecraft electronics TINY particles of dust pose a more serious risk to satellites than huge lumps of space junk, according to a team of British scientists who have analyzed damage to solar panels from the Hubble Space Telescope. Most bits of space junk, even pieces as small as a few centimeters across, are tracked by the US Air Force. This allows any spacecraft that is on a collision course to be moved out of the way. But some particles are too small to be tracked by radar. This week scientists from the Open University and Oxford Brookes University will tell the Hypervelocity Impact Symposium in Galveston, Texas, that these particles, just a few micrometres across, pose a serious threat. They believe that dust grains vaporize on impact, creating a hot, conducting plasma that can induce currents that seriously perturb electronic systems on spacecraft, potentially rendering them helpless. ; One way to assess the risks posed by space dust is to examine the tiny craters which pepper the surface of space shuttles after orbital missions. But this is often impossible, as telltale residues in the craters are usually burnt away as the shuttle re-enters the Earth's atmosphere. So the British team examined solar cells removed from Hubble at the same time as its famously defective mirror was replaced. The panels were brought down in the shuttle's hold, so the residues remained intact. The researchers found that the craters (see inset Photograph) on the solar cells contain enough residue from impacting dust particles to identify the culprits, say Giles Graham at the OU and Anton Kearsley at Oxford Brookes. They found traces of iron, nickel and magnesium in the craters. This suggests the damaging particles were asteroids or comet debris. 'We also found residues of aluminum and titanium from space debris in the smallest craters,' says Graham. Neil McBride of the OU says the findings highlight the risk posed to satellites by the numerous particles of space dust orbiting Earth: 'Natural particles can be traveling up to seven times faster than space debris particles, which means they produce over a thousand times more plasma on impact.' The warning is timely, because on 17 November the Earth will pass through the tail of Comet Tempel-Tuttle, and the debris from it will rain down on Earth-producing the display better knows as the Leonid meteors. Next year's Leonid shower is expected to be the strongest since 1966. The stream of debris will be rich in micro-particles. 'There's a very serious risk to satellites,' says David Asher of the Armagh Observatory in Northern Ireland. McBride points out that a satellite called Olympus suffered an electrical malfunction at the peak of the 1993 Perseid meteor shower, which eventually led to its loss. 'We can't say that it was definitely hit by a particle, but it seems likely,' he says. Author: Eugenie Samuel http:// www.newscientist, COM issue 11 Nov 2003

Fig 5

Deep Space Galaxies, Hubble Space Telescope

Galaxies, galaxies everywhere—as far as NASA's Hubble Space Telescope can see. This view of nearly 10,000 galaxies is the deepest visible-light image of the cosmos. Called the Hubble Ultra Deep Field, this galaxy-studded view represents a "deep" core sample of the universe, cutting across billions of light-years. The snapshot includes galaxies of various ages, sizes, shapes, and colors. The smallest, reddest galaxies, about 100, may be among the most distant known, existing when the universe was just 800 million years old. The nearest galaxies—the larger, brighter, well-defined spirals and ellipticals—thrived about 1 billion years ago, when the cosmos was 13 billion years old. In vibrant contrast to the rich harvest of classic spiral and elliptical galaxies, there is a zoo of oddball galaxies littering the field. Some look like toothpicks; others like links on a bracelet. A few appear to be interacting. These oddball galaxies chronicle a period when the universe was younger and more chaotic. Order and structure were just beginning to emerge. The Ultra Deep Field observations, taken by the Advanced Camera for Surveys, represent a narrow, deep view of the cosmos. Peering into the Ultra Deep Field is like looking through an eight-foot-long soda straw. In ground-based photographs, the patch of sky in which the galaxies reside (just one-tenth the diameter of the full Moon) is largely empty. Located in the constellation, Fornax, the region is so empty that only a handful of stars within the Milky Way galaxy can be seen in the image. In this image, blue and green correspond to colors that can be seen by the human eye, such as hot, young, blue stars and the glow of Sun-like stars in the disks of galaxies. Red represents near-infrared light, which is invisible to the human eye, such as the red glow of dustenshrouded galaxies. The image required 800 exposures taken over the course of 400 Hubble orbits around Earth. The total amount of exposure time was 11.3 days, taken between Sept. 24, 2003 and Jan. 16, 2004.

Credit: "http://www.nasa.gov/", "http://hubble.esa.int/", S. Beckwith ("http://www.stsci.edu/") and the HUDF Team

Image Type: Astronomical

STScI-PRC2004-07a

Ions tossed into space from every star in the universe are wending their way through space and will be encountered constantly. If you doubt the existence of ions from stars just realize where our Aurora Borealis comes from and those are only ions from our own sun. We must now talk of things generally unknown by the general population. A Black hole is so dense that nothing can escape from the gravitational pull of a Black Hole, including light. There is an event horizon generated by the gravity of a Black Hole that exists around the Black Hole. The distance or height of the event horizon is proportional to the mass of the Black Hole. Hence, if we took our spaceship to a Black Hole outside of its event horizon and dispatched a visible lighted probe from our ship toward the Black Hole the probe would be seen until it gets to the event horizon where it will seem to disappear. In reality the light from the probe is being pulled into the Black hole and is no longer directed toward you; thus you can't see it after it passes the event horizon.

Black Holes have other properties that must be discussed. They emit particles from them back into space. How does a particle get away from an infinitely dense black hole that holds onto everything in its vicinity? That's not too hard to answer, the particle escapes the black hole by leaving at faster than the speed of light. It will be faster than the speed of light till it clears the event horizon where it slows down to light speed shortly after clearing the event horizon. More over, black holes spew these particles out with great frequency adding to the dust, ions, energies, nebulas et al—and other particles in space (herein after referred to as (Space air, SA). OK, so there is a bunch of debris in space, in fact space is loaded with it from many sources, let me postulate that just as air molecules hold flying vehicles to the speed of sound as a possible maximum and water molecules limits the speed of vehicles in water that space dust limits the speed of vehicles in space to the speed of light, because that is the maximum speed that some of the space dust can reach. implied in this postulation is the fact that the concentration of space dust will vary throughout the universe, hence the speed of light will vary as an inverse proportion to the density of these dust populations. One can conclude that space dust will be at maximum density close to a star with a planet population such as our own solar system. Light, being pure energy will still move at the speed of light and not meet to much opposition, but space ships attempting to travel at light speed or greater will meet enough opposition to destroy themselves if traveling to fast for the conditions of the road Light, being pure energy will still move at the speed of light and not meet to much opposition, but space ships attempting to travel at light speed or greater will meet enough opposition to destroy themselves if traveling to fast for the conditions of the road. We have previously stated that a warp bubble is virtually impossible to construct for it is thought that all of the available energy in the universe would be needed in order to make a warp bubble large enough to cover one small electron. It is now time to come to some conclusions about the subject we have been perusing all along. Does and has the universe always existed or did God step in and create it? We will pursue that conclusion after we make some observations about what we think we have learned thus far.

Chapter 6

The World Around Us.

We humans are an astounding race of beings. Sentient, learning, caring, loving race of beings that usually tends to get lost, in fantasy and daydreams, while reveling in those fantasy worlds, thereby losing track of what is happening all about us. We claim to be reasonable beings who use logic to explain all things that happen in both the real and imaginary world. We are also stubborn enough to cling faithfully onto our observations even if there is a fragment of doubt within us to contradict what we are stating. All things aside, the reality of this occurs when our sun goes runs out of hydrogen, we are going with it, unless we have a new home somewhere in this universe. Now it seems that traveling at light speed or greater is not obtainable in the immediate future. This indicates that we have problems to solve before we go off into space beyond our solar system. Can they be solved, certainly, for I hold the greatest of hope in the brilliance of the mind of man. We will solve these speed problems and hop off into the galaxy. Inter galactic travel seems unlikely even long term however because of the huge distances that must be traveled not just through our own galaxy (200,000 light years) but throughout the space between galaxies (millions of light years). We would probably have to travel in a condition of Stasis for generations but that presents a whole new set of problems we haven't touched upon (who will do the procreating & who will teach the young generation after generation). I would think that traveling throughout our galaxy ought to solve the problem at least until the black hole in the center of our galaxy decides to gather in all the stars & planets in the galaxy, through gravitational force, just as it now keeps our galaxy intact. This indicates that beyond the boundaries of our universe there is nothing out there. Please remember in our introduction we stated that two groups of thought pertinent to the universe were conflicting. If the universe has been forever and will forever blow apart, reach equilibrium, draw back together again and blow once again ad infinitum. If the universe was created by God then this is not an endless cycle but instead a one time event.

Chapter 7

Future of the Universe

We lack data to support the theory that the universe has expanded beyond the limit that gravity can pull it back together again. We also lack knowledge of what happened at t=0 of the big bang minus a short period of time. In point of fact we can not even begin to realize this infinitely dense mass right at the time of the big band or know why it went bang. Our lack of knowledge may just be due to our inability to spot alternatives. Space, as in the space enclosed within our galaxy is filled with sub-atomic particles, ions, photons which are thought to be both energy and matter with mass. This matter flying around in space comes from space dust, ions shed by stars and debris left over from the big bang and other things. Most of it is fairly harmless and is filtered out by our planets atmosphere. However, if you were flying in space at warp speeds a collision with a particle could destroy a ship. Please remember that it could be traveling at the speed of light. If you are also traveling near light speed the resultant closing speed of the collision would be warp 2. A small particle could punch a hole through your ship and anyone else in its path. Shields won't do you any good for the force necessary to deflect this particle would require more energy than exists in the universe. I, non-the-less believe that we will solve these problems and find a way. Now, we've covered our traveling into space. Let's talk bit about people from other worlds visiting us? Interesting for we are now looking directly into the face of ET. There is no reason to think that ET didn't have the same problems to overcome that we now face. From what we now know and can foresee it appears that the problem of warp travel is going to remain a mystery. Even if you allow that ET has solved the problems pertinent to warp travel & warp bubbles he is going to be limited by the fact that maximum speed is a function of space air and its accumulated debris and therefore will be forced to have maximum warp he can travel lest he goes to fast and the debris punctures his vessel. So lets assume that he can accomplish warp 11 and goes at speeds 2048 times faster than the speed of light. This would indicate that he would need about 100 years just to cross our galaxy. Intergalactic travel having upwards of millions of light years in-between galaxies present insurmountable distances for even ET to be trying to overcome. For this reason I find it hard to believe that man has ever been visited by aliens. It just takes to long to get here and if you must return, to whom do you report your journey for very likely all that saw you leave will be dead themselves. Before moving on to the last chapter we must explore our universe. Please don't get me wrong, I do not claim the universe for mankind for I do not believe that we will ever be out there to claim other galactic planets. This does not conclude that no other life exists in the universe for the pure mathematical possibilities of their existence are overwhelming and their existence must be taken as a foregone conclusion. Let's explore the size of the universe for a few sentences. We humans have always believed that our universe went up in a Big Bang about some15 billion years ago, hence with a multidirectional expansion outward we come to the conclusion that the universe is 30 billion light years in diameter. Simple, 15 billion years times 1 light years distance per year. As we look into the skies we try to find all the matter that the scientific community says has to be there in order to have a Big Bang and we come up short. Not enough subatomic particles, not enough dark matter, not enough of everything combined to make up the matter for a Big Bang. In fact we seem to be short about 90% of the stuff that makes a Big Bang. Surely one can not think that God took some matter in order to create parallel universes, Not that he couldn't if he wanted to but it doesn't seem to likely. Let me try to speculate as to where all the missing matter could be for the real reason for stating that God made the universe is to prove that the existing universe is expanding faster than gravity will ever be able to pull it back together. This brings us to the problem of realizing where the real universe is, not just the universe that we can sense with IR telescopes or ordinary telescopes including the Hubble telescope. None of our instruments has the range to spot things that are moving away from us faster than the speed of light. There are black holes at the edge of the universe, then the particles emitted by these black holes could escape the black hole and roam on into the universe unhindered in their flight away from the black hole. Let's take that speculation one step further and imagine that at t=0 when the boom went Bang that 90% of the matter that exploded got pushed away from the explosion at greater than the speed of light, after all there is nothing in space to impede or limit its velocity. If this is a possibility coupled with debris cast off by black holes at the edge of the universe then the universe is not 30 billion light years in diameter but it would be closer to 60 billion light years in diameter. Now, since one of the factors disallowing victory to either the universe forever people or the "God did it" people was that neither the "God did it" people nor the scientists could reach a solidified opinion on the expanding universe. If the universe was only 30 billion light years in diameter and only a few percent of the matter was found to make up a big bang then we could not come to a conclusion that the universe would either stop or expand forever. If, however, the universe is larger than we thought and looks like it will expand forever then we can conclude that the universe came with only one Big Bang and that occurred about 15 billion years ago. If this is the conclusion we come to then we must ask the question "Who made the Big Bang?" To that there is only one answer, God made the Big Bang, therefore God exists. I suppose the next logical question would be,

"why did God make it go Bang?" We are not going to state any conjecture about that

question. Suffice to say that in making the statement "God exists" one has chased the

mind of God and if we can agree that the conclusion is true and God exists would it

not be prudent to drop the matter there rather than opting to probe any deeper in the

Mind of God. I'm sure that what we have touched in God's mind is only a perfunctory

tap on the tip of His brain. Do we dare to attempt to probe deeper into His mind or

is there a limit to what He will allow for a disturbance. Obviously what God knows

is infinitely more than what man will ever know and perhaps this is the way it should

be. The conclusion of God's existence alone should be enough to appease man and

probably tickle God's sense of humor simply because we know he now exists. What

then can we accomplish in addition to knowing of God's existence? I would hazard a

guess as to what God views when he looks down on his creation as shown as follows:

Fig 6   God's View

Epilog

If the argument in this work is correct and all premises are found to be acceptable, then the conclusion is correct and God created the universe. There can be no mathematical proof for this postulation for, in order to accomplish proof, we would have to expand our mathematical base into what is now called the unification theory. This math would be a combination of our current quantum mechanics with relativity and a new math. Quantum mechanics explains all that happens in the subatomic world. The two forms of math are not interchangeable. Relativity explains all that happens in the world above the sub atomic level. We have no math for what happens at the instant of the Big Bang or any time previous to the Big Bang but if we are smart enough to find the mathematical answer to this period of time we will probably encompass the three theories into one that we hope will work in all three worlds. This new math (the unification theory) has been worked on for year mathematicians and physicists have worked diligently on this problem. It is probably at least as complicated as Relativity and it could be a lot more complicated. To facilitate understanding of the postulation of this work, we will then use only logic which is another mathematical basis for coming to the truth. The basis of the two factions claims are that the universe at its currently known expansion has not developed enough to tell yet if it is going to continue to expand, stop expanding and crunch back to a singularity and go bang again or hit a point of stability and neither expand nor collapse. The opponents simply argue that the universe has always been expanding the contracting and infinitum. If we could show that the universe is larger than we think it is then the argument is ended and we can conclude that God created the universe because it only went Bang once and won't do it again. If we can accept the theory that at the time of the Big Bang a large percentage of the matter of the Big Bang got ejected into space at greater than the speed of light, then our conclusion is correct. After all space outside the compacted matter of the Big Bang was completely void of energy or matter and there was nothing to prevent the matter of the Big Bang from flying off at greater than light speeds. All current theories state that nothing exists outside the boundary of the existing universe but empty space, therefore nothing would stop matter in a complete void from going faster than light speeds given the impetus to garner enough velocity to do so. If this is the case then, the universe is possibly double the size necessary to have it be

larger than we now think it is. That would make it large enough to state

that it will never stop expanding, hence it only went Bang once and will

never do it again. The conclusion is that God created the universe as well

as causing the Big Bang. As a side line we can state that traveling through

space is going to be harder than we thought due to the Relativity Inference

on Matter in motion approaching the speed of light if for no other reason

than that we can't create warp speeds. The other inference is that what we

consider to be the boundaries of our universe are wrong. They are set to close

in to what actually exists. This makes sense, since we can not observe light

from particles that are traveling faster than light away from us which make

our observations impossible. We have constantly been probing for answers

to cosmology questions and have found that we frequently have to change

the way we think in order to accomplish our research goals. I believe that it

is so, for the answer to the Unification theory that we now seek but to get

there we may have to adjust our present knowledge base into more logic

conforming to our observations. all in all we have sought the mind of God

since the beginning of reasoning. This is a noble chase and I can't believe

that God cares if we touch his mind once in a while but we should never lose

touch with reality and infer that we are a smart as God. Does this conclusion change the life of men? Even though the ultimate Armageddon fate of the Earth won't happen for billions of years, we, as a race of beings, must begin soon to find answers to some questions pertinent to space travel lest the time comes and we are not prepared. Nothing is to come easy to man and finding an answer to the final Armageddon dilemma will prove a benefit to mankind overall if we can find the solution before we need it. Will we continue to probe the questions of the reality of the universe? If we don't we will continuously encounter problems that will curtail our overall progress and place detriments in the course of mankind as time passes. It would be remiss of me not to put things into their proper dimension. We humans tend to find our hats growing a bit too tight from time to time and in this case it could be caused because we just discovered that God exists. If man were to fly in from outside the universe to pay us a visit, his journey would be a very long one. That's not to say that God can't be any place he chooses to be in an instant but in relation to how man would have to travel. Let me offer the following scenario: Man would have to cover the entire universe and as stated the universe is 90 billion trillion miles across at this time (but growing every second). Figure 8 depicts a section of the Universe in a radius around our galaxy. If man started at the edge of the galaxy He would have to travel almost 21 billion trillion miles just to get to the focal point of that radius.

Our galaxy is located at the focal point of the arc.

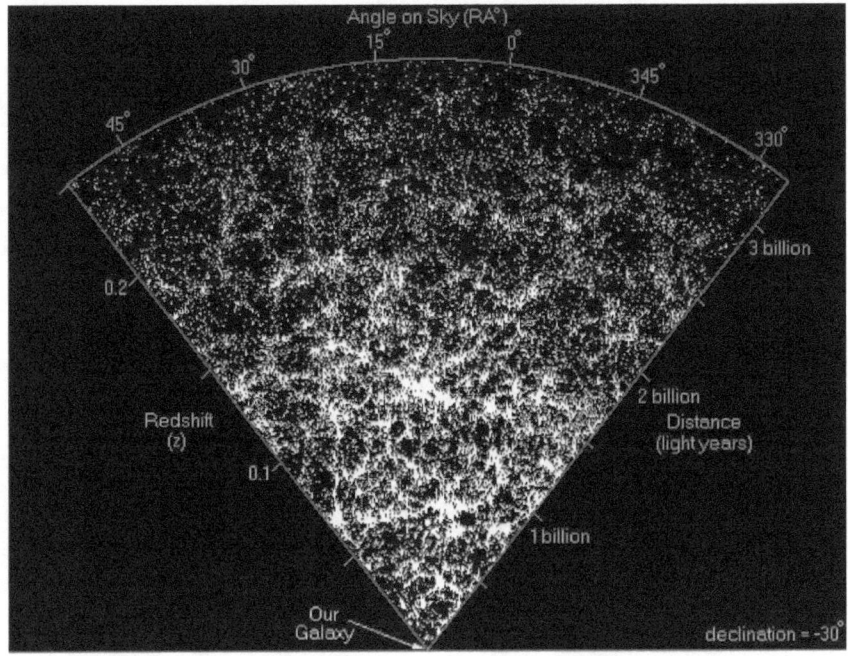

Fig 7

Arc of the Universe with The Milky Way at the focal point (bottom of the arc) It must be pointed out that each dot of light in Fig 8 is a galaxy in the universe. It would be difficult to spot which one is the galaxy you want to visit since dots of light are similar at a distance. Navigation within the universe would be difficult since there is no North, South, East or West for all galaxies are in movement with respect to each other and moving at different speeds. The first leg of our journey is 21 Billion, Trillion miles long and we still are not close to the Milky Way. As we approach the Milky Way Galaxy from a distance of 60 million, billion we can begin to see the galaxy but it's still 60,000,000,000,000,000 miles off. Figure 9 depicts a view from that distance.

Fig 8
Milky Way At 60,000,000,000,000,000 miles

The point of the arrow of Figure 9 shows the location of the Milky Way. One can see just how hard navigating the universe can be. To get a little closer we shall move up to approx. 8 light years as shown in Figure 10.

Fig 9   Milky Way at 8 light years

Still to hard to pick out so let's move to a closer range of 5 light years!

Fig 10

Milky Way from 5 light years away

It is beginning to look like a spiral galaxy now but finding which one of the 200 difficult. Let's move in closer. Our sun is seen at the point of a small white arrow at about 7 O'clock on the picture.

Fig 11

Milky Way from 2 light years away

Let's move in a little closer to only a few million miles as seen in Fig 13!

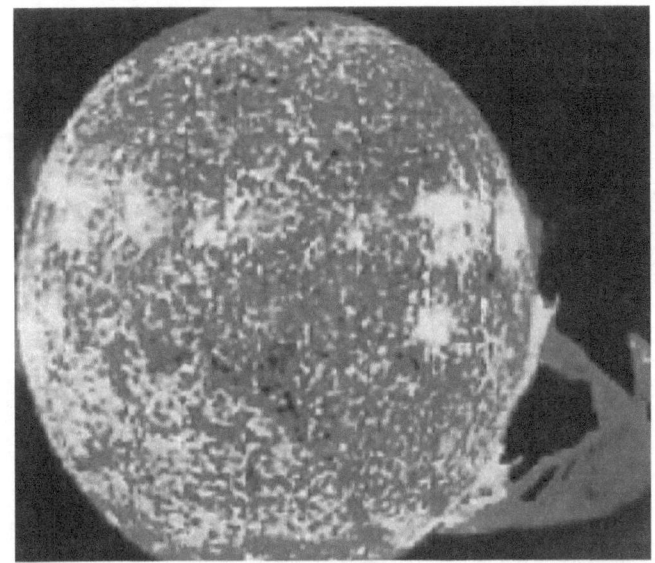

Fig 12

Our Sun from a few million miles away

Locate our solar system rotating around the Sun as per Fig 13

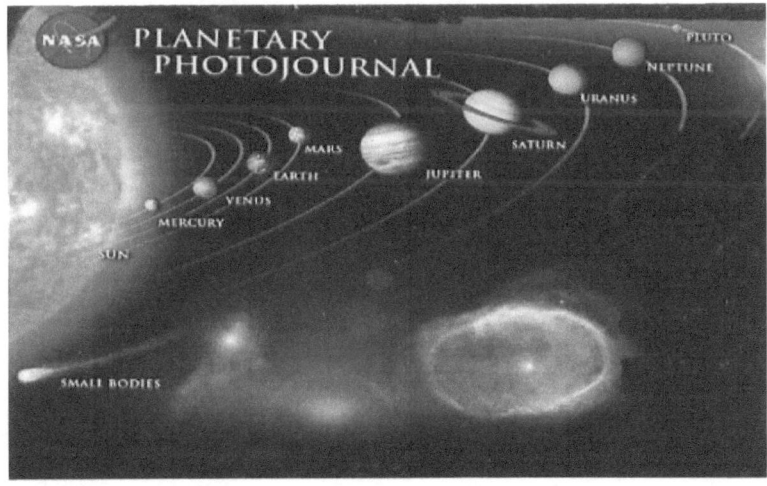

Fig 13

Our Sun and its Planets

Moving toward Earth one will see the following in Figure 15!

Fig 14   Earth

Moving still closer to our goal we can observe the California coast. from a few hundred thousand feet up! See Fig 15!

Figure 15

Marina Del Rey in California

approaching our original target we land on a California beach! See Figure 16!

On the Beach in California Fig 16

We have taken this trip from outside the universe to the Earth to show two things. One, the distance involved in making such a journey covering miles in astronomical numbers and, two just how insignificant mans size is in relation to the totality of the Universe. Man is just about six feet tall and is about .000000000000000000000001% the size of the universe. In other words, we are insignificant by comparison to the universe. Another reason for bringing up these differences is the fact that man often when seeking answers to questions will turn up more new questions than answers. An example of one of these questions concerns black holes. The big bang was started by an exploding singularity that contained all the matter thought to be in the entire universe. That singularity must have been infinitely dense. Black holes, on the other hand, emit particles and energy constantly. This is an indication that they are not infinitely dense. Consider a regular atom that has a nucleus of protons and neutrons about the size of a golf ball. The electron in it's first orbit would be the size of a BB and be located about a block away from the nucleus. The subsequent orbits of the electrons would be another block out and away from the first orbit and the other orbits also each be about a block further than the previous orbits. When a black hole is formed, the atoms collapse from their orbits in order to compact the atom. Could there be degrees of compactness of this collapse such that the electrons only fall back 50-95% from their original orbit, and head toward the center of the atom. This compacting of atoms would lead to varying densities of black holes which would allow particles and energy to escape through the spaces from the black hole if that particle had enough velocity to escape the event horizon of the black hole. This indicates that there is a difference between black holes and singularities with the singularity being infinitely dense and the black hole being less than infinitely dense. We shall end our journey at this point so that we can go out and find some more answers or at least some more questions. Opponents will argue "how made the singularity go bang" In response to this let me simply point out the power of a being powerful enough to make miracles happen. This being must have the power to make simple matter explode. If not by his will, then allow that he could flood the core of the singularity with anti-matter which he also created and the singularity simply exploded from the reaction of the matter and anti-matter, scattering the matter in the outburst deep into the heavens. Since there was nothing in space to slow the propelled matter in its space travel, that matter left the explosion at greater than the speed of light. The extremities of the area this scattered matter occupied at the extremities of its moving is the outer envelope of our universe. It is obviously a size that God can handle easily but man becomes overwhelmed by this immense size, hence, what to God is a manageable size is almost more than our human comprehension.

The universe operates according to God's rules and will therefore always be a manageable size to Him even though to us it is rushing towards infinity never again to be pulled back to a singularity to become a big bang again. The last point is in considering where the missing 90% of the matter went that was attributed to the totality of matter needed to make the big bang. Assuming that 10-20 % of the big bang matter scattered away from ground zero at greater that the speed of light would make the size of the universe twice the size we now assume it is if that initial 20% escaped from the big bang point at twice the speed of light and there was nothing in space at that time to slow it down. The light from those stars formed from that 20% of the initial matter would never be discovered by science because it would be traveling away from us faster than the speed of light. The remaining 70% of missing matter would have been consumed in the matter-anti-matter explosion that occurred in the big bang collision of matter and anti-matter causing the big bang. That matter had to be someplace and matter and energy are interchangeable in science. Hence, the missing 90% of the matter that made up the big bang is either out of sight far away from us, never to be seen by us and was also used in the explosion that was the big bang. We face a future of limited space travel because we can not travel faster than the speed of light. If the human race is still around when the sun burns off all its hydrogen, we will then face Armageddon as the sun envelops the first four planets that are in its orbit. We humans have proven that we constantly find new ways to annihilate ourselves and thus may not be around to participate in the consumption of the Earth by the Sun. If this is the case, either we learn to change our behavior lest we don't get through Armageddon in either the biblical sense or the scientific definition. The choice is mankind's to make, peace or face extinction. Let us go back now to the expansion of the universe, recent experiments have yielded a surprise, the universe is still expanding. This means one of two things. The universe was not measured at the diameter but at the radius. This fact alone increases the size of the universe from 90 billion-trillion miles to 180 billion-trillion miles from side to side. Further the universe is not expanding at the speed of light but, instead, at faster than the speed of light, which makes it even larger than 180 billion-trillion miles across. No one knows when the universe may slow down to the speed of light but it has not happened in the last 15 billion years. Scientists have, however, decided that this expansion is to great to ever have the universe stop expanding and crunch back to the tiny atom where it began. What them becomes of our planet and the universe. Our planet is going to be alright as long as mankind does not kill themselves as well as all life on the planet. Doing this does not guarantee safety for us. There is still a threat of a cosmic disturbance causing. Another sizable asteroid or comet could collide with the Earth and again cause global extinction. The Bible could be right and Armageddon could occur. For certain, when the sun runs out of hydrogen, it will expand its corona encompass the first three planets in it's orbit, which included the Earth. The atmosphere will be blown away, all water will evaporate into space and all life on Earth will be burned into extinction. All that will be left is a rather large rock floating in space.

Let us consider this thought of there being parallel universes, an infinite number no less. This was produced by the introduction of string theory. This explains that there is an infinite number of universes in parallel dimensions to ours. They are connected by worm holes to each other. Unfortunately, an emanate scientist, Prof Stephen Hawking, proposed the idea of an infinite number of universes in his "The Information Paradox". Years later he published a composition that disputed his own theory and the business of an infinite number of universes should have disappeared. If there is only one universe, and we have answered the question, "How did the Universe Begin?" And, we discovered that GOD created her "Big Bang", them we must ask, "How will the Universe end". To begin with, we know that the universe is already 15 billion years old and is still expanding and will continue to expand even further, quite possibly for another 100 billion years. That of course is only a guess but what we can establish is that the Universe is running from energy. That energy is derived from the energy of the Big Bang. As it continues to expand that energy will escape into the infinity of space and dissipate until there s no measurable energy. Those Hugh black holes at the center of all galaxies are already dispersing energy into space and will continue to do so until the black hole no longer exits. AS the entire Universe burns out, there will be no energy left to form stars and galaxies and the Universe will be a bunch of wildly spaced lumps of rock neither reflecting nor emitting energy. Drawn to nothing because gravity, which is a weak force to begin with will not be strong enough to draw matter toward other matter because of distance. Thus, when all the black holes have dissipated, and all the galaxies have drifted far into space, there will be no energy left to form new stars and galaxies and the Universe will dissipate into dark space, lifeless and totally sterile. People will say, "Why worry about 100 billion years from now? You do not have to worry about what may happen in 100 billion years. But in 5 billion years our sun will run out of hydrogen. It will expand to include in it's corona, the first three planets of our solar system. The Earth is the third planet. The Earth will become a cinder without air, water or any life. If we are not off this planet and traveled to another and settled on this new planet, will become a cinder like all of life on Earth.

We do have a problem that we do not know how to solve at this time. How to travel long distances, such as a billion light years, without discovering warp drive. Science now tell us that if we took all the energy In the entire universe, we may be able to make warp bubble large enough to shield an atom. This is a far cry from shielding a space ship of enormous size, and so we will perish when our sun dies and becomes a red dwarf. Can we learn enough to develop a small power warp bubble. Perhaps but that is a highly speculative proposal. What we must do is have faith that people will, perhaps, find the energy to create a warp bubble or find a worm hole. This faith is the only remedy that can be seen to solve our problem but we have quite a bit of time before this solution will be needed. We must keep the faith for if we slow down our efforts, disaster will eventually overcome us, that is a certainly. There are two other aspects that must be approached in order to complete this text. Dark Matter and Dark Force. They are relevant to the conclusion of this text. Dark matter exists as a matter from the Big Bang and is matter that we can not see nor prove that it exists but has as it's purpose for Being is that fact that it holds the galaxies in their place and keep them from falling apart as spinning into infinity as they would to if allowed to spin and move as they wished. It is estimated that between the matter observed in the universe including the billion galaxies, Plus the dark matter equals only 27% of the matter from the Big Bang. The other 73% exists as Dark Force, again undetectable but acting as a reverse agent to gravity by filling in the voids in space and acting as a repellent fighting against gravity. Scientists have assumed that around 5 billion years after the Big Bang that the forces of gravity and the Dark Force ceased to be in competition and the Dark Force took over to dominate the Universe. This is why scientists. When they had the tools to measure the Universe, were surprised to find out that the universe was accelerating in it's expansion instead of slowing down as gravity was supposed to do in it's effect. We have, for several years tried to find and prove the existence of Dark Matter and the Dark Force and have failed to find anything that may resemble its existence.

Little matter whether found or not, the fate on the Universe is still not changing. As the Universe will expand until it drifts so far into the Universe that all the Black Holes will expend their energy and disappear into the void of space. Galaxies will be overcome by the Dark Force and spin so far into the void that their energy, even after super novas Will dissipate into empty space. Eventually, all active matter will cool down and the universe will cease to exist, becoming icy rocks floating in space. All that will be left will be icy balls so far apart that the Universe will not exist as it is now known. There will be no life left in the Universe.

Bibliography

Cover—Hubble Space Telescope, The brightest star in the sky. Eta Carinae

No 1—dirbe123_2 p6dec-Cobe

No 2—dirbe123_2 p6dec

No 3—HST Galaxy M51

No 4—HST 1100 @ Anglo Australian Observatory

No 5—HST-WFPC2 Distant Galaxy Cluster Around 3C-324

No 6—HST WFPC2 Distant Galaxy around 3C 324

No 7—www.motodom.com/galaxy.htm

No 8—www.motodom.com/galaxy.htm

No 9—Microsoft clip art

No 10—www.galacticimages.com

No 11—www.galacticimages.com

No 12—www.milkyway.com

No 13—Microsoft clip art

No 14—Microsoft clip art

No 15—Microsoft clip art

No 16—www.globeexplorer.com

No 17—www.globeexplorer.com

www.ingramcontent.com/pod-product-compliance
Lightning Source LLC
Chambersburg PA
CBHW021244280526
45784CB00005B/2228